The Ocean World of Jacques Cousteau

Oasis in Space

ANGUS AND ROBERTSON

TO THE MEMORY OF JAMES DUGAN

First published in 1973 by

ANGUS & ROBERTSON (U.K.) LTD.
2 Fisher Street, London, WC1
102 Glover Street, Cremorne, Sydney
107 Elizabeth Street. Melbourne

First published in the United States of America
by The World Publishing Company

ISBN 0 207 95481 X

Printed in Great Britain by W. S. Cowell Ltd.

"This little lionfish nuzzling the reef for food has been caught and classified many times by marine biologists in the past 200 years. We know more about his anatomy than about his habits. We do know that he carries venom in some of his dorsal spines that could be extremely toxic to man, but that he only uses this for emergency defense. What does he sense with his little brain, what does he feel within his little body? Is all his life motivated only by simple impulses, like hunger or sexual drive? Is he more than a beautiful mechanical toy? I know there will be no answer to such questions in my lifetime. But the mystery of life is as inspiring as music or poetry."

Contents

In the oceans have been tested all conceivable BEHAVIOR PATTERNS (Chapter VII). Some of these patterns are protective, some destructive: a dolphin could very well survive alone, but if isolated from his companions he prefers to die. Nature displays every combination.

Most of the time the oceans are in a state of "armed truce" interrupted by brief periods of hunting and violence. The complexity of the relationships has created an intricate, fragile, but stable *balance*: INTERDEPENDENCE (Chapter VIII).

At the turn of the century, when Prince Albert I of Monaco brought back to his oceanographic vessel the first catch of hundreds of abyssal fish from 20,000 feet of depth; when Jacques Piccard and Don Walsh in their historic dive in the bathyscaph *Trieste* witnessed life at the bottom of the Mariana Trench; when Albert Falco steered diving saucers on the continental shelf—they were just beginning to understand the SEASONS AND PROVINCES OF THE OCEAN (Chapter IX).

But in the sunny layers of the tropical sea, innumerable small polyps have built structures that dwarf the most ambitious man-made constructions: THE PHARAOHS OF THE SEA (Chapter X).

Man is turning today to the RICHES FROM THE SEA (Chapter XI). But some of these resources have limitations: overfishing today depletes reserves: offshore oil is not inexhaustible; whaling brings about extinction of the largest animal ever to have existed.

Man's RETURN TO THE SEA (Chapter XII) is dramatically illustrated by the historical telephone conversation in 1965 between Philippe Cousteau and Scott Carpenter.

Because of overfishing, mismanagement, ruthless destruction, and pollution, today THE SEA IS IN DANGER (Chapter XIII). Obviously, an individual man is intelligent enough to understand that he is not a dinosaur, but *as a species* he may not be responsible enough to avoid the fate of the dinosaur.

Frightened by his concrete jungles, man turns to the oceans for leisure and escape, only to find that they are dying from his carelessness. Man takes off for the conquest of space, only to find that the solar system is a dustbin of dead celestial bodies. The truth is that man is alone—a lonely, pulsating, thinking creature—in the universe. Alone on board his Spaceship Earth, an OASIS IN SPACE (Chapter XIV), his only cousin is the dolphin, the only creature with which he can hope one day to communicate.

Introduction: If the Oceans Should Die

It is a shocking paradox that at the precise moment in history when man is arriving at an understanding of the sea he should also have to face the question above. Just now, in our generation, when after many thousands of years of ignorance and superstitions man is at last beginning to learn about managing and exploiting the vast resources of 70 per cent of earth's surface, he finds himself in a race against time to rescue it from his own spoilations.

If the oceans of earth should die—that is, if life in the oceans were suddenly, somehow to come to an end—it would be the final as well as the greatest catastrophe in the troublous story of man and the other animals and plants with whom man shares this planet.

To begin with, bereft of life the ocean would at once foul. Such a colossal stench born of decaying organic matter would rise from the ocean wasteland that it would of itself suffice to drive man back from all coastal regions. Far harsher consequences would soon follow. The ocean is earth's principal buffer, keeping balances intact between the different salts and gases of which our lives are composed and on which they depend. With no life in the seas the carbon dioxide content of the atmosphere would set forth on an inexorable climb. When this CO_2 level passed a certain point the "greenhouse effect" would come into operation: heat radiating outwards from earth to space would be trapped beneath the stratosphere, shooting up sea-level temperatures. At both North and South Poles the icecaps would melt. The oceans would rise perhaps 100 feet in a small number of years. All earth's major cities would be inundated. To avoid drowning one third of the world's population would be compelled to flee to hills and mountains, hills and mountains unready to receive these people, unable to produce enough food for them. Among many other consequences of the death of the oceans, the surface would become coated with a thick film of dead organic matter, affecting the evaporation process, reducing rain, and starting global drought and famine.

Even now the disaster is only entering its terminal phase. Packed together on various highlands, starving, subject to bizarre storms and diseases, with families and societies totally disrupted, what is left of mankind begins to suffer from anoxia—lack of oxygen—caused by the extinction of plankton algae and the reduction of land vegetation. Pinned in the narrow belt between dead seas and sterile mountain-slopes man coughs out his last moments in unutterable agony. Maybe thirty to fifty years after the ocean has died the last man on earth takes his own last breath. Organic life on the planet is reduced to bacteria and a few scavenger insects.

Why begin a work on the subject I love most in the world with this nightmare? Because the ocean can die—and because we want to make sure that it doesn't. Man exists only because his home-planet, Earth, is the one celestial body we know of where life is at all possible. And life is possible on earth because earth is a "water planet"—water being a compound itself probably as rare in the universe as life, perhaps even synonymous with life. Water is not only rare, not only infinitely precious—it is peculiar, with many oddities in its physical and chemical make-ups. It is out of this unique nature of water, interacting with the dynamics of the world "water system," of which the sun and the ocean are the motors, that life originated. The ocean is life.

This is why we must change our attitudes toward the ocean. We must regard it as no longer a mystery, a menace, something so vast and invulnerable that we need not concern ourselves with it, a dark and sinister abode of secrets and wonders. Nor do we want to follow the methods of the first scientists who sailed and searched the seas to compile lists: lists of mammals, lists of sea-birds, of jelly-fish, of temperatures, of currents, of migratory patterns. Instead we want to explore the themes of the ocean's existence—how it moves and breathes, how it experiences dramas and seasons, how it nourishes its hosts of living things, how it harmonizes the physical and biological rhythms of the whole earth, what hurts it and what feeds it—not least of all, what are its stories.

I dedicate these books to water and the life which depends on water—and to the mother of waters: the Ocean.

Jacques-Yves Cousteau

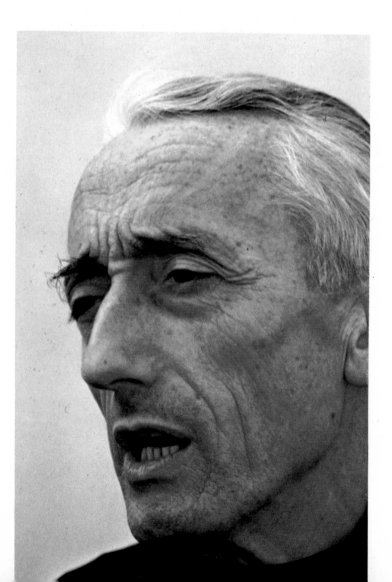

Chapter I. Water–The Essence of Life

Of all the earth's resources, water is the most precious. Without it no plant or animal could have evolved.

Where did the water come from that makes life possible? The most widely accepted theory proposes that water in the form of vapor was one of the compounds present in very small amounts in the vast gaseous cloud of hydrogen and helium from which our sun evolved. The sun became a separate body without using up all the matter in the cosmic

> **"Every drop of water on earth when earth was formed remains on our earth today."**

cloud. What remained whirled through space and after millions of years became the nine planets of our solar system. The water from the original cloud, according to this theory, became a part of each planet, but was not necessarily evenly distributed. The amount of water present and the form it took—solid, liquid, or vapor—depended on the planet's mass and its distance from the sun. The moon, for example, is too small to retain an atmosphere or water vapor.

Through the happiest of accidents, the earth is the right size and distance from the sun to permit water to exist in all three forms. It is in its liquid form that water is essential to the life process. And it is in liquid form that water is rare in the universe. Close to the sun the heat is so intense that water is vaporized. Far from the sun it is so cold that water remains permanently frozen. Of the other planets, only Mars, our second nearest neighbor, is in the narrow temperature band in which water can exist in its three states. Thus, extraterrestrial life in our solar system is deemed possible only on earth and maybe Mars. Whether life, in fact, exists on Mars we do not yet know.

Every drop of water on earth when earth was formed remains on our planet today. It has since gone many times around the globe; water is constantly moving. It is pulled from the oceans by the heat of the sun and moved along in clouds by winds. It may fall from the sky as rain, snow, sleet, or hail. It may fall in any latitude from Arctic to Antarctic. It will stay for ages in icecaps and glaciers, or return promptly to the sea through storm sewers and rivers. Almost all of the water on earth today is in the oceans. Life itself is thought to have begun there. No one is sure exactly how. But at an unknown time, probably 3 billions of years ago, some fortuitous circumstance made it possible for chemical substances dissolved in the warm broth of the sea to combine with each other to form molecules with the unique ability of reproducing themselves. From those first living molecules all life has descended.

A / Ice. Captain Cousteau and his team preparing to film the walrus in the semi-frozen waters of Alaska.

B / Fresh water. Deep-sea divers in a fresh-water grotto.

C / Oceans. Earth as seen from space.

D / Clusters of clouds. Yet another form water takes.

Chapter II. A Man Is Born—A Dolphin Is Born

A man is born. The most elaborate creature on earth is frail, he has no natural weapons for attack or defense. For years he will need the constant care of his mother, the protection of his father. All his life his body will have to fight gravity. He is a visual creature, sight being his most crucial sense. He is gifted with the ability to laugh, to play, to make love in all seasons. He is a social animal, but is often aggressively violent toward his own kind. He has access to the knowledge accumulated by his ancestors thanks to written or electronic storage of information. This baby faces a destiny that can range from the miserable condition of an urban slum dweller to the fascinating life of an outer-space explorer.

A dolphin is born. The most elaborate creature in the sea also needs the care of his mother for about a year. He is weightless in the sea so that most of his muscles can be devoted to propulsion. His life-expectancy is half that of his human counterpart. He is an acoustical creature, capable of "seeing with his ears" much farther than with his eyes. He plays most of his time, has an active sex life all year round. His social rules are severe, but he is not aggressive. He probably communicates by a modulated language, but we don't know yet to what extent he can store and transmit information. This baby's destiny is one of an easy, happy life, sometimes complicated by parasites, rarely endangered by natural enemies.

The Dolphin—A Marine Playboy?

The dolphin's life is a little like that of a human playboy's. He doesn't live as long; 30 is probably the maximum age for most species of dolphin. But he moves into the interesting business of life much more swiftly than the man. He has a gestation period of 12 months and a lactation period of another 9 months—somewhat similar to the human rhythm. As with man the dolphin protects its young from danger. To ward off shark attacks a group of adults may encircle the young as a barrier.

A dolphin is sexually mature at eight years —and the female can give birth once a year after this for another 14 years or so. In their free natural habitat dolphins tend to breed in spring and summer. Early morning seems to be the favorite mating hour. But the animal is sexually active all year round. Sexual experimentation of all kinds, body-rubbing, mouthing, biting, flipper-touching form a big part of the dolphin's life. And he has plenty

> "The dolphin prefers to play. When all else failed to amuse him, the dolphin invented body-surfing."

of opportunity for it. For he is an awesomely efficient swimming machine, equipped with acoustic range-finders and a capacious brain, who can count on providing himself a daily diet without wasting much time on it.

With a body beautifully streamlined to minimize resistance while coursing through the oceans, and a powerful tail that makes it possible for him to stand two-thirds of his body's length out of the water for seconds at a time ("sculling"), the dolphin can if required swim at a velocity of 20 miles an hour almost indefinitely. He is also an expert diver. His blood can absorb more oxygen than man's can and he has additional oxygen stored in muscle reservoirs. By slowing his heart beat and restricting the flow of blood except to the brain he makes the best use of this oxygen in his dives. His tissues can tolerate an oxygen debt and his brain is less sensitive to CO_2 than ours. His lungs are undamaged by complete collapse. As a result the dolphin can easily dive deeper than 300 feet and remain underwater for more than five minutes. Some have reached depths of 1000 feet—where the naked human diver rarely dives to 100 feet.

These remarkable adaptations and the dolphin's physical appearance point up the long years that have elapsed since dolphins separated from their terrestrial ancestors and embarked on their zoological adventure in the undersea—to become one of the most specialized and efficient marine creatures that has ever existed. But this all sounds like work. The dolphin prefers to play. When all else failed to amuse him, ages ago, the dolphin invented body-surfing—and in many parts of the world can be seen rollicking in the breakers in groups of four or more. With the advent of the steamship the dolphin discovered another game — riding the bow waves. Perfectly calculating the speed of the ship, the contours of waves, winds, and currents, one or more dolphins take up a position in front of a fast-moving boat and, with next to no expenditure of energy on their part, hitch a ride. Let man do the work!

Unlike us, the dolphin has found no need to develop his territory by constructing huge habitation complexes. He is content to play, perpetrate his species, and take from nature only what he needs. Such an animal has no enemy in the sea except for his cousin the killer whale. And, alas, man, who cheats—with net, harpoon, and gun!

Man-Dolphin Life Cycle

1. GESTATION — 9 MOS. / 12 MOS.

2. AGE AT SEXUAL MATURITY — 10-12 YRS. / 8 YRS.

3. AVERAGE LONGEVITY — 65-85 YRS. / 25-35 YRS.

4. AGE WHEN WEANED — 6-12 MOS. / 6-18 MOS.

5. LENGTH OF PARENTAL PROTECTION — 8-12 YRS. / 2 YRS.

6. HOW LONG TOTALLY HELPLESS — 1 YR. / FEW HOURS

7. TOOTH ERUPTION — 6 MOS. / FEW WEEKS

Chapter III. Cradle of Life

Ours is a watery planet. The oceans, which cover almost three quarters of the earth's surface, contain some 330 million cubic miles of water. Dissolved in that huge liquid expanse are a variety of salts and minerals

> "A 'dry water cycle' took place millions of times before the earth cooled enough to permit the water to remain and accumulate in cracks and depressions."

and the gases necessary to the teeming and almost unimaginably varied life in the sea. Dissolved oxygen is breathed by aquatic animals. Dissolved carbon dioxide is extracted by marine plants and used by them to produce food.

There was no sea during the earth's infancy while the planet's crust was forming and cooling. At an early stage water released from our hot planet vaporized and rose to envelop it in a vast cloud. This water cooled, condensed and fell to earth during almost permanent thunderstorms, only to be revaporized and rise again. This "dry water cycle" took place millions of times before the earth cooled enough to permit the water to remain and accumulate in cracks and depressions.

If we trace the long infinitely slow process of evolution backwards, we would begin with the two highest creatures yet to evolve: man, the most intelligent creature on land; and the dolphin, the most intelligent creature in the sea. Both man and dolphin belong to the same animal grouping—both are warm-blooded, air-breathing mammals, vertebrates that nurse their young. Man and dolphin are both descendants of the early mammals that roamed the land—sea creatures who for a time lived on the continents and then returned to the sea. These prehistoric mammals were heirs of cold-blooded vertebrates who had evolved from the first fish that managed to crawl from the water and adjust to life on land. This transition involved adjusting to the full force of gravity, not felt underwater, and obtaining oxygen from air instead of water. Those air-breathing and walking fish in turn had evolved, through many steps and millions of years, from primitive animals without backbones who were distant descendants of one-celled animals. But even before the animal world of the oceans could evolve animal life and plant life had become differentiated. Organisms that acquired the capacity to use energy from the sun to manufacture food for themselves out of the chemicals dissolved in the sea became specialized as plants.

> "It is one of man's illusions that the oceans and land masses of the world are eternal."

It is one of man's illusions that the oceans and land masses of the world are eternal and that he himself is the ultimate end product of billions of years of evolution. Though there are reasons to believe that the rate at which new forms of life are produced has substantially slowed down, the earth remains in a constant state of flux. In the process of evolution which continues from moment to moment, the human species interferes dramatically. Today, his influence is tragically destructive. But we can hope that Science and Reason will reverse the trend. In any case, the world of the future will surely be as different from the world of today as today's world is from that of the past.

Solid Earth v. Liquid Ocean

Were the crust of earth to be levelled—with great mountain ranges like the Himalayas and ocean abysses like the Mariana Trench evened out — no land at all would show above the surface of the sea. Earth would be covered by a uniform sheet of water—more than 10,000 feet deep! So overwhelming the ocean seems to be. The earth is truly a water planet. Yet there is another way of looking at it, perhaps more appropriate when thinking of man. All the water on earth, fresh and salt, makes up only a little more than one tenth of 1 per cent of the volume of earth itself.

In this perspective one understands why we must conserve and cherish this limited and vital resource. Water makes our earth livable. It is, in fact, life itself. Water modifies our climate by storing heat and redistributing it over the earth through ocean currents and atmospheric circulation. Water is the universal solvent — allowing more substances to be dissolved in it than any other compound. In man water is the vehicle which transports vital materials throughout the body, making possible life's complex biochemical reactions. Water is indeed a rare and precious gift!

17

Water Is Precious

Water, along with sunlight and oxygen, is one of the prime ingredients that make life on earth possible. Without water life as we know it just can't be. So water is one of our most precious assets. Deserts are virtually waterless and virtually lifeless too. Oceans, on the other hand, are virtually all water, and the oceans teem with life.

A / Mars. The canyons and terraces suggest the presence of water—is there life there too?

B / The sea. A lively school of jacks.

C / Fresh-water marsh. Abounding with plant and animal life.

D / The Moon. Without water—dead dust and rock.

E / Mangrove swamp. Where marine life begins in the tropics.

F / Mountains. Water in both its solid and liquid states—depending on altitude.

G / Desert. Waterless, and thus nearly lifeless.

H / Grasslands. An abundance of water—an abundance of greenery.

I / Polar icecap. Life near and about water even in extremes of temperature.

Calypso at Dawn

"It is early morning. I am alone on the deck of the research ship *Calypso*. The sea is flat and grayish-blue. The air is hot and sultry. There is hardly any breeze. Overhead, heavy dark clouds are silently piling up into billowing pyramids. We are moving through a belt of water slightly north of the equator known as the doldrums. In the days of wind-driven sailing ships mariners feared these waters. The doldrums is the region where the trade winds never blow. It is where ships used to becalm and where supplies would run out. But I hear the steady throbbing of *Calypso's* engines. I know this ship will pass through the doldrums without mishap.

"What is it about the sea, I wonder as I stare into its depths, that has always lured man on to explore its secrets? I think of the great voyagers of the past—Ericson, Magellan, Vasco da Gama, Francis Drake. Their daring navigation told us much about the size and shape of the oceans of the world. Today we know the configuration of the surface of the sea with an accuracy they could not have dreamed of.

"It is the underwater world we now explore. We are looking there for clues to our own distant past, for information about the formation of our planet and the origin of life. *Calypso* is crammed with equipment to aid us in that search. Our underwater observation chamber enables us to watch and photograph marine creatures wherever we venture. Divers going down to depths of 200 feet for a firsthand look at the ocean's habitats enter and leave the water through our midships' diving-well. Our echo-sounders help us map the ocean floor. Other research vessels have drills that bore into the earth's crust and bring up cylindrical samples of that ancient material. We are like detectives trying to unravel what happened millions and billions of years ago.

"Somewhere on the ocean floor are bits and pieces telling of the story of this planet's beginning. It is in the oceans, I believe, rather than the skies that we will finally learn how earth and life formed. And the sea, which gave birth to us all, will shelter us once again. Looking down at the water I can imagine man-made islands of the future floating offshore and other settlements rising from the ocean floor. Before man leaves earth for an-

> "Where did this bountiful planet come from? Are there other planets anything like ours in other solar systems? Do they too, if they exist, have comparable forms of life?"

other home I am sure he will retrace his steps through the sea.

"Raindrops begin to fall. They too come from the ocean, I remember, as I climb the steps to *Calypso's* bridge. It is at peaceful moments like this that I ask myself the questions I am sure man has been asking himself without answers for almost as long as he has inhabited the earth. When you raise your eyes to the sky, you inquire—where did this bountiful planet come from? Are there other planets anything like ours in other solar systems? Do they too, if they exist, have comparable forms of life? Or life we cannot imagine? Where did the spark of life on this planet come from?"

Calypso—originally a minesweeper built in America which served in the Royal Navy and after World War II was converted to one of the great ladies of ocean science: tough and tireless.

▲ HUMPBACK WHALE

Chemical Content of Human Blood and Ocean

Na⁺

140 - 150

0
100
200
300
400
470
500
600

K⁺

4.2 - 6.2

1
2
3
4
5
6
7
8
9
10.0

Ca²⁺

2.5 - 5.2

0
1
2
3
4
5
6
7
8
9
10
10.2

Cl⁻

103 - 116

0
100
200
300
400
500
548
600

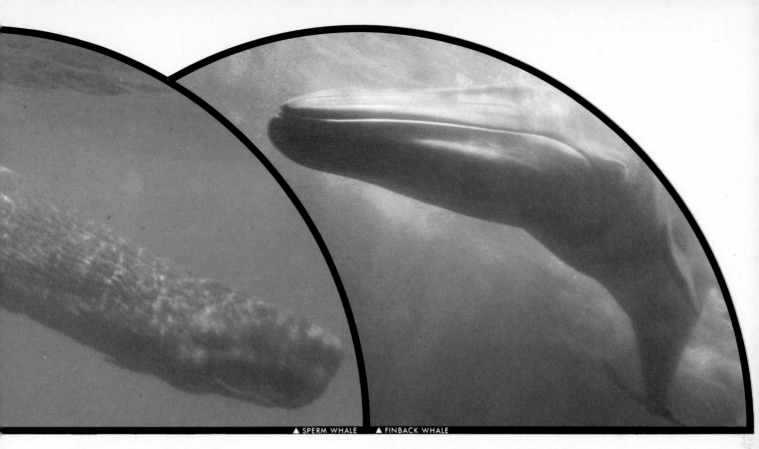

▲ SPERM WHALE   ▲ FINBACK WHALE

From Land to Sea

Whales possess reduced bones representing four limbs, hips, and an articulated neck. Unlike fish, they lack bones in the caudal fluke and dorsal fin. The story of whales most probably began with a primitive land-dwelling mammal 125 million years ago. This creature of the distant past may have found it advantageous to search for food in marshes and estuaries rather than compete on land for mice or lizards. At one stage our primitive whale may have looked similar to a seal. Millions of years of trial and error culminated in the contemporary cetacean, highly adapted to the marine environment but helpless on land. In addition to the whale's other physical features he now possesses an underwater communication system.

In the diagram to the left, chemical symbols are used. "Na" stands for sodium, "K" stands for potassium, "Ca" for calcium, and "Cl" for chlorine. The four pictures show the relative amounts of these elements in sea water and in human blood.

From Sea to Blood

Today we are pretty sure that life originated in the sea. But around the turn of the century the provocative suggestion was made that man might be carrying a proof of this around in his own circulatory system. It was noticed that there is a similarity in the salt contents of human blood and sea water—especially in the electrolytes most crucial to life: sodium, potassium, calcium, and chlorine. The salts in sea water are at about three times the concentration you find them in blood, but the ratios between these ions are intriguingly alike. We know that sea water has been growing saltier through the eons. Could human blood be an echo of the primitive ocean that was less salty, the ocean as it existed when vertebrates first appeared and established their circulatory systems? The theory has recently come in for hard knocks because it has been demonstrated that even the primitive ocean at the time vertebrates emerged was twice as salty as blood.

23

Half-and-Half Existence— Land and Sea

Walruses live around the world—but only in the Arctic. Although there's only one species recognized by scientists, they divide them into an Atlantic and a Pacific race. The Atlantic race lives along the seacoast, often on icefloes, in the Canadian Arctic, Greenland, and in European Arctic waters east to Siberia. The Pacific walrus comes from Alaskan and Siberian Arctic coastal waters. Walruses dig clams out of the ocean floor down to a depth of 240 feet with tusks which are overgrown canine teeth that may reach a length of 20 inches. Then they apparently suck the clams out of their shells.

Male walruses get to be 12 or 14 feet long and may weigh up to 3300 pounds. Females get to 10 or 12 feet in length and a ton in weight. They give birth to single pups every two or three years in the spring. The pups weigh over 100 pounds at birth. Walruses live for about 30 years. Their biggest enemy is man, mostly the Eskimo who lives off the walrus's three-inch layer of blubber, flesh, skin, bone, and tusks. But they are protected from overhunting now. A few years ago a census revealed about 25,000 Pacific walruses and about 30,000 Atlantic walruses left in the world.

Between Two Worlds

"In between two worlds — ours and the abyssal depths — a solitary nautilus has come up to the 200-foot level to feed. I can scarcely believe it! Here is the elusive survivor of hundreds of millions of years beneath the sea! Studied mostly in shell and fossil form, and briefly in captivity, here at last in the wild is the *living* animal. A living fossil! Shy of light, he slowly moves away towards his secret home in the depths. Jet-propelled, he swings rhythmically like the pendulum of an eternal clock. The nautilus has ticked off many ages of evolution. It has seen men returning to the sea as it saw the dolphin return to the sea."

◄ *The common jack or crevalle is a fast-swimming, tireless hunter.*

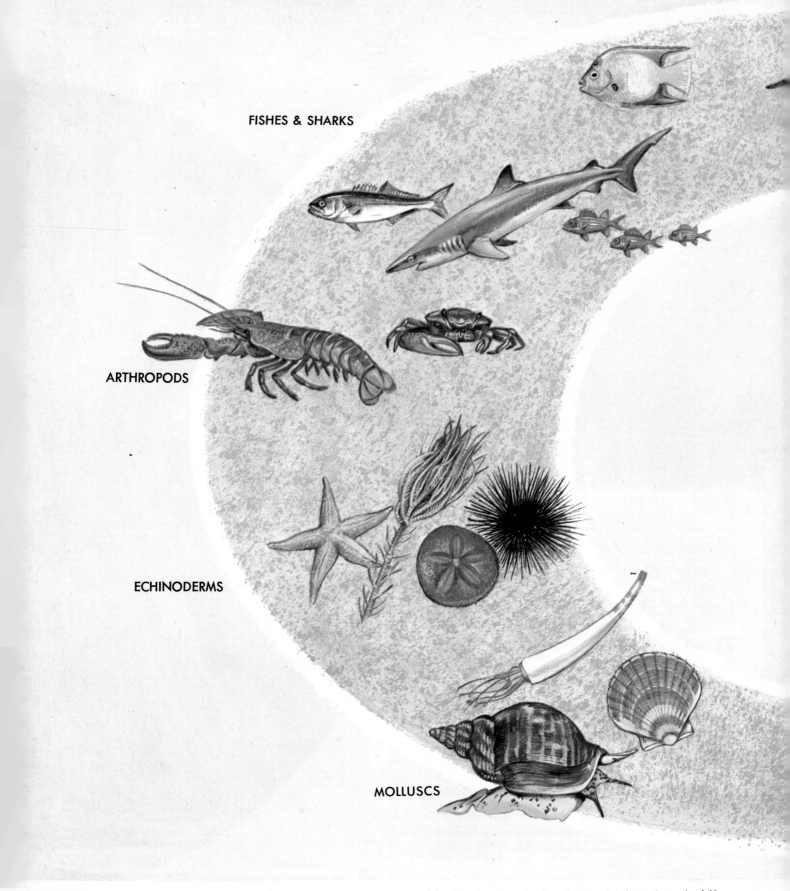

FISHES & SHARKS

ARTHROPODS

ECHINODERMS

MOLLUSCS

The Animal World

All life in the Animal World is related. All life, the most primitive as well as the most

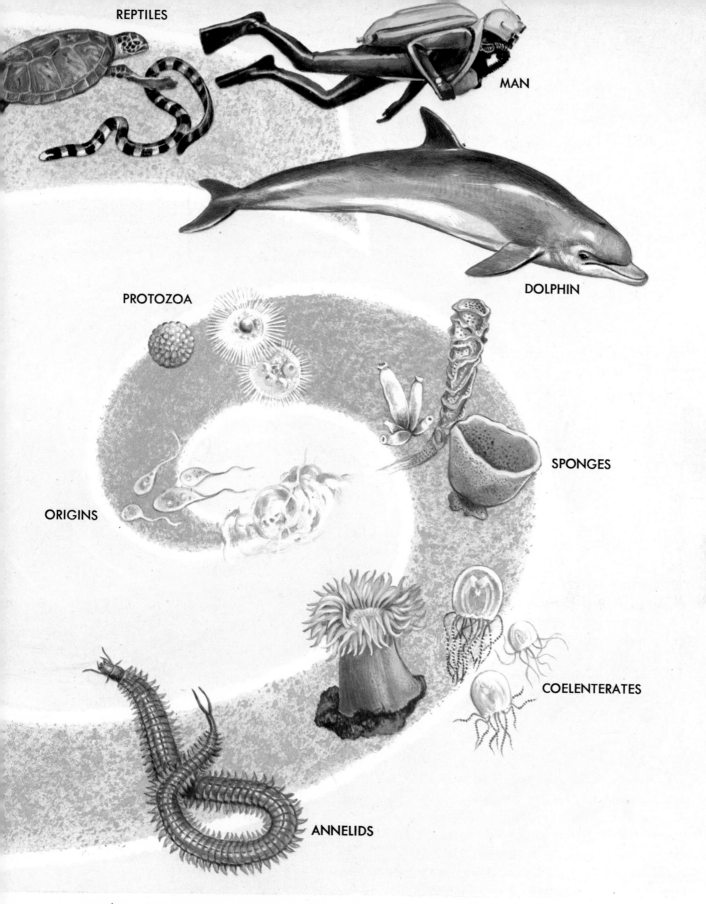

REPTILES

MAN

DOLPHIN

PROTOZOA

ORIGINS

SPONGES

COELENTERATES

ANNELIDS

complex, stems from a common origin. All living things depend upon the same heredi-tary material, the common thread that binds animals and plants together.

Chapter IV. Essential Drives

The fundamental law of the animal world in the sea, as on land, is: survive. To that end animals must eat and must procreate. Only if it gets enough food can an animal grow to maturity. And only when mature can animals reproduce and ensure survival. Sea animals feed on each other in a set order

> "Half of 1 per cent of the solar energy falling on the ocean remains in it due to ocean plants."

that moves upward from the lowest to the highest. For all but the swiftest and most powerful, the general rule is eat and be eaten. But with great ingenuity and ferocious energy and tenacity, animals at all levels instinctively protect themselves so the act of reproduction can take place.

Man, though an animal himself, subject to hunger, thirst, the need for shelter, the urge for sexual gratification, and the will to live, has learned through centuries of civilization to tame his instincts and defer his pleasures. Many of man's basic needs are today readily fulfilled. It is thus not easy for modern man to imagine the blind power of the basic drives that move the animal world through its cycles. Other animals act instinctively. It is only in extreme circumstances, when his very life is at stake, that modern man can appreciate the force of the drive to survive.

Survival for the individual is, of course, finally impossible. Animals and men must die. And nature is unsentimental about when and how. But for all the indifference and brutality of life in the wild, it contains little unnecessary slaughter. Animals kill to eat. Dolphins, which like man, have leisure time, spend it peacefully. Man, however, the most intelligent of creatures, is now a threat to himself and to all other life on the earth.

The food chain which keeps the Animal World going is energized by the sun. Half of 1 per cent of the solar energy falling on the ocean remains in it due to ocean plants, mainly the unicellular algae in the plankton, which are the "primary producers." Tiny herbivores eat these plants, tiny primary carnivores eat the herbivores, larger secondary carnivores may eat both primary carnivores and herbivores. And so it goes—with all life of the sea playing its part: the sperm whale chases giant squids through dark waters nearly a mile down; the mako shark slashes at a school of mackerel; the northern fur seal waits for the evening rise of lantern fish; almost imperceptible organic particles slowly drift down to the glass sponges and brittle stars of the abyss. But if the simpler forms spend almost all their lives searching for as much food as they can get, on the contrary, the highly evolved creatures can sat-

> "If simpler forms spend almost all their lives searching for as much food as they can get, highly evolved creatures can devote time to more complicated occupations—play and social relationships, affection and love."

isfy their needs easily and thus devote time to more complicated occupations—play and social relationships, even affection or love.

The Food Chain. Paradoxes abound in the food chain in the sea as on land. Like the elephant on the dry land, the mightiest animals of the sea—the blue whale among the mammals and the whale shark among the fishes—are not the bloodthirsty predators their size and strength would seem to entitle them to be. Huge as they are, they are gentle browsers among the smallest creatures of the sea, the minuscule inhabitants of the plankton layers.

MAN FISHES
THE SEAS
FOR FOOD

SUNLIGHT—PHOTOSYNTHESIS

BACTERIA
DECOMPOSES
ORGANIC MATERIAL
FORMING
NUTRIENTS

PHYTOPLANKTON

ZOOPLANKTON

SMALL FISH

TUNA

SHARK

DEAD MATTER AND
ANIMAL WASTES ARE
CONVERTED INTO
USEFUL CHEMICALS
BY BACTERIA

NUTRIENTS RETURN TO THE SURFACE • FEEDING AND FORMING PLANKTON

Home In the Harem

You are looking in on a small part of a colony of northern fur seals deployed over one of their breeding grounds. The central institutions of seal society are the harems, organized in late spring by mature bulls, five years old or older, who climb ashore at the rookeries after ten months of "batching" it in the open seas. There is a good deal of shoving and heaving, and some serious fighting, before the subadult males and the aging ones have been chased out of the main camp and each breeding bull has staked out a private territorial claim. At this point the females swim ashore—most of them in the last stages of their pregnancies from the previous season. Each female finds her bull and moves into his harem. How large a harem depends on many factors, but it can be sizeable; a big male northern fur seal can control as many as 100 cows. The pups are born, and within a few weeks the adult seals have

> "At the height of the season a male northern fur seal can mate three times an hour and can maintain a rate of one act of love an hour for periods lasting up to three weeks."

mated again after a courtship of perhaps only a few moments. (The bull seal is no refined lover.) Cows tend to be recruited into harems at about five years of age, but it is known that a few four-year-olds breed outside the colonies—adolescents have created problems since the beginning of time. Twenty to thirty days after the mating period the harems break up and the bulls depart, leaving behind their females and suckling pups, not to see them again until the next spring. One of the reasons for such a "sexist" arrangement may lie in the size-discrepancy between male and female seals. Very much bigger and stronger than their mates, the males may constitute a danger. The males pay for these prerogatives. It is only rarely that a male seal attains the age of twenty-five, while females can live as much as ten years longer. Still, it is all very similar to an Arabian Nights establishment—which is why the appellation "harem" was applied to the seal colony in the first place. At the height of the season a male northern fur seal can mate three times an hour, and can main-

tain a rate of one act of love an hour for periods lasting up to three weeks. Fast and big and intelligent, seals take pretty good care of themselves in the open sea. Only the faster and bigger and more intelligent killer whale is too much for them. But in their summer rookeries—overwhelmed with the pressing business of establishing territories, fighting off interlopers and eager bachelors, breeding, mating, scolding, suckling, and preparing for the long winter ahead—the seal is absolutely defenseless against the clubs and guns of the human seal-hunter. So those which carried valuable furs on their backs were all but exterminated from the planet. Perhaps man has stopped just in time, for with a few tragic exceptions the seal family —including walruses, sea-lions, and elephant seals—now seem to be fairly secure.

Attack and Defense

A / Shark. Rows of sharp teeth line the mouth of the shark.

B / Sea snake. Its venom is lethal, but a sea snake has such a small mouth it can bite a diver only in places like the fold of skin between the thumb and index finger.

C / Walrus. The sturdy tusks of the walrus grow up to two feet in length.

D / Blue crab. This crustacean uses its oversized claws for courtship gestures as well as defense.

E / Sea urchin. The sharp, brittle spines of the sea urchin present a formidable defense.

F / Surgeonfish. Knife-like spines are found on each side of the surgeonfish near the tail.

G / Orca. The "killer whale," as it is mistakenly known, attacks with its large, sharp teeth.

H / Moray eel. The razor-sharp teeth in the mouth of the moray eel give this fish a menacing look.

I / Lionfish. The fearless lionfish is equipped with a row of 18 needle-like dorsal spines.

J / Tridacna clam. The heavy shells (bivalves) of the tridacna clam clamp firmly together.

Comparing Senses of Man and Dolphin

Man is a visual animal. The dolphin is an acoustic animal. Each has what he needs most in his own world. From a hilltop or promontory, on a clear day, the normal man can see as far as 100 miles. But on the "clearest day" in the sea the eye cannot reach much farther than 100 feet. Even crystal-clear water is actually a fog so far as the organs of sight are concerned. In returning to the sea the dolphin had to adapt to this

suggesting that the entire structure could be involved in responding to sound. In practicing echolocation the dolphin emits a fascinating range of noises—clicks, tone signals, whistles—both for bouncing ahead of him or communicating with his fellows. His hearing range is very much larger than man's—going up to 150,000 cycles to man's top limit of about 12,000. Of the dolphin's other senses probably the visual is next in import-

undersea limitation. He was able to compensate thanks to the hypertrophy of his auditory faculty, modifying greatly the structures of the internal and external ears in order to receive and interpret a wide range of waterborne vibrations. All this not only vastly improved the animal's hearing

> "From a hilltop, on a clear day, a man can see 100 miles.
> But on the clearest days in the sea the eye cannot reach much farther than 100 feet."

underwater but refined it for the purposes of echolocation—the sonar-like process with which the dolphin homes in on objects as distant as a mile away that he wants to inspect—perhaps to eat. In a sense the dolphin's whole skull may be an "ear," because his ear bones are connected to it in a fashion

ance, though he cannot match man in this faculty. Dolphins in marinelands see well enough to leap out of the water for the fish their trainers dangle before them. He must also experience some sensations of his skin, otherwise he would not make love as he does, but of course he has no hands—and therefore little sense of touch. As the dolphin possesses no olfactory nerves, we must assume that he has no sense of smell, but it is possible that some hitherto unidentified area of cranial tissue may deliver to his brain some of the aspects of smell. Nor does the dolphin seem to possess taste buds or salivary glands. What he does have is a brain comparable to that of man.

Second sight. Although humans and dolphins can see about the same distance underwater, the dolphin has an extra sense, a "sonar," which permits him to perceive events precisely and over much greater distances.

The Potential for Learning

Intelligence, meaning the capacity to apply what one has already learned to new circumstances, is generally thought to be a function of brain size. As a rule, the larger the brain the greater the potential for intelligence. However, size alone does not guarantee superior intelligence. The elephant has a bigger brain than man but is not more intelligent.

Brain size is correlated with body size. As an animal increases in overall size, its nervous system must likewise increase in size or the animal will become unable to meet the challenges of its environment. One reason the dinosaur became extinct is that its body outgrew the ability of its nervous system to function efficiently. Given an adequate brain size and presupposing that animals continue to learn as long as they live, how much an animal learns will depend on how long it lives. Man, as one of the longest-lived of animals, can, over a lifetime, learn more than most other creatures. And one thing he has already learned is how to increase his own lifespan. More than a century ago, in 1850, the average life expectancy in the United States was about 41 years. Today, largely through expanded medical knowledge of formerly fatal childhood diseases and the development of life-saving drugs, the average life expectancy has risen to 70 years.

The dolphin, having a large brain and living to about 30 years, acquires much information in a lifetime. Since dolphins make many kinds of noises—grunts, calls, clicks, squeaks, barks, and whistles—some observers presume that these sounds are the means by which the animals communicate with each other. By communication the observers do not mean calls for help or cries of pleasure, which are emitted by many animals, but the actual transfer of information through "language." Doubt is cast, however, on whether dolphins have a true language and even on their general level of intelligence by the fact that the creatures are caught over and over again in the same regions of the world in the nets fishermen put out for tuna. One would suppose that if dolphins truly had both language and high intelligence one of them would have warned the others long ago and all dolphins would have learned to avoid these dangerous waters.

In any case, the dolphin is surely more intelligent than any other marine creature. Recent laboratory experiments shed some light on this intelligence. Some researchers in Hawaii have taught dolphins to create —that is, to initiate new behavior. The experiment began by teaching a dolphin a trick and rewarding him for it. Upon returning a day later the trainer would not give a reward for that same trick. Eventually the dolphin would give up and start to play, whereupon the first new act in his play would be rewarded. After a few days the dolphin would begin each session not with an old trick but with a new behavior. It was concluded that the dolphin understood that he must "think up" something new to please the trainer and earn his prize.

> "The dolphin is more intelligent than any other marine creature. It is capable of creating—that is, of initiating new behavior."

Lifespan and civilization: The longer the expected lifespan the more information the animal can gather and use. In turn, as in the case of man, that knowledge has helped him increase his longevity.

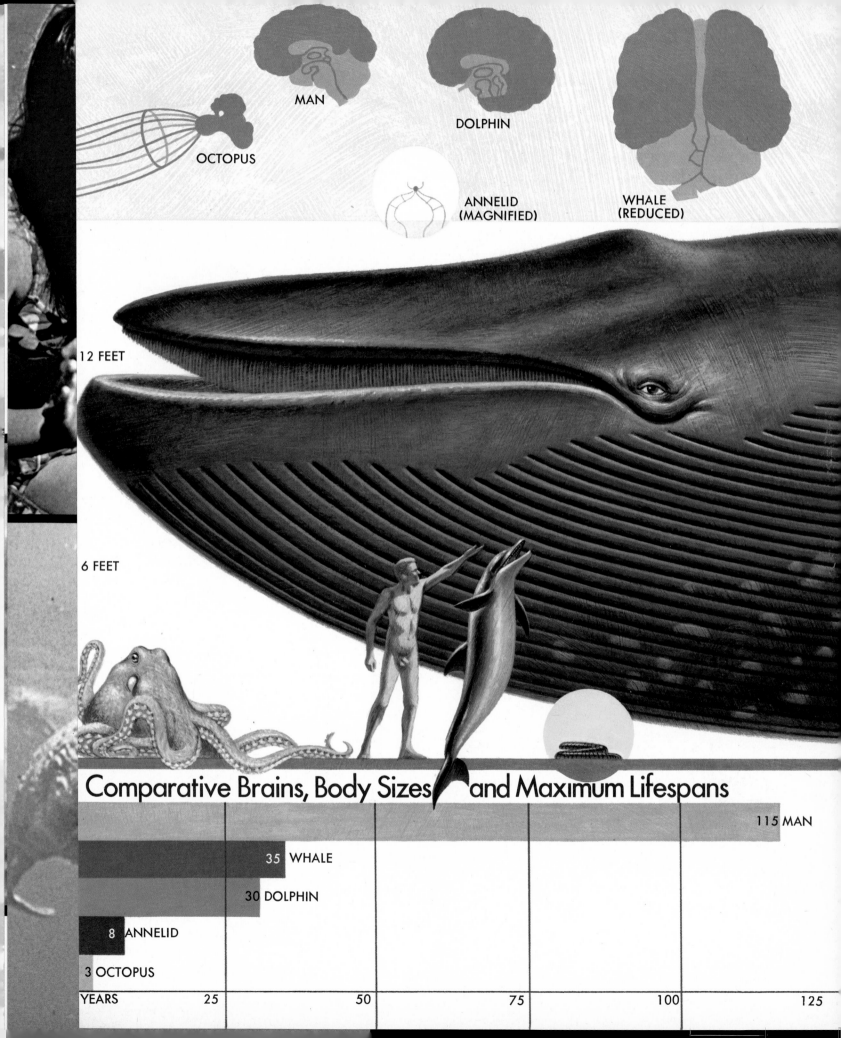

OCTOPUS

MAN

DOLPHIN

ANNELID
(MAGNIFIED)

WHALE
(REDUCED)

12 FEET

6 FEET

Comparative Brains, Body Sizes and Maximum Lifespans

		115 MAN
35 WHALE		
30 DOLPHIN		
8 ANNELID		
3 OCTOPUS		

| YEARS | 25 | 50 | 75 | 100 | 125 |

Ways Animals Adapt

A / Moorish idol. The extended fins of the moorish idol provide stability, its shortened body maneuverability.

B / Toadfish. This slow-moving bottom-dweller is well camouflaged and hides in tin cans and other debris.

C / Jellyfish. Only 1 per cent of the jellyfish's body is the muscle tissue it needs to move through water in lazy jet propulsion.

D / Grouper. The enlarged head of the grouper acts as a powerful suction pump for feeding.

E | Regal angelfish. The extended snout of this spectacularly colored shallow-water fish is perfectly fitted to pluck small invertebrates from their hiding places.

F | The spiny puffer. The scales of this puffer, or porcupine fish, have evolved into spines. When the puffer feels threatened, it is able to inflate itself by inhaling either air or water. The resulting thorny ball lacks appeal to most predators.

G | One spot butterfly fish. Instead of camouflage for protection, these fish fool their predators. The deceptive eyespot directs an attacker away from the head.

H | Wolf eel. This ferocious looking fish has a mouth filled with fangs and molars.

▲ A   ▼ B

The Arctic Scene

A / Polar bears. *These huge animals are a source of both danger and food. There are now fewer than 10,000 in the world—mostly in Canada.*

B / Arctic fox. *This beautiful animal has one major natural enemy: the polar bear.*

C / Eskimo family. *A typical Eskimo family relies on polar bears, walruses, whales, caribou, and reindeer for their food and clothing. Armed with modern weapons, the Eskimo are beginning to endanger some of the species of their native regions.*

D / Penguins. *These birds, natives of the Antarctic, do not yet suffer the predations of man.*

E / Narwhal. *This whale with a single tusk is possibly one of the inspirations for the fabled unicorn.*

F / Gyrfalcon. *A predatory bird that lives off small animals and other birds.*

G / Seal. *An animal that has too frequently been useful to man for its lustrous coat.*

H / Wolf eel. *It is sometimes as long as nine months after birth before this eel begins to eat.*

▲C

▲F

▲D ▼E

▲G ▼H

At the Center—Calcium and Carbon

The beautiful reef on the opposite page has a story to tell—one of the most important in the book of the oceans. For like all other reefs this one is composed largely of calcium carbonate, and the cycling and recycling of calcium carbonate through the undersea world lie at its heart. On land calcium carbonate is found mainly in the form of limestone, deposits of which take up about 22 per cent of earth's land area. Rain and wind patiently gnaw away at these deposits, washing them into the sea in solution as calcium bicarbonate at the rate of 500 million metric tons each year. So vast is this input that the sea would soon be saturated with calcium bicarbonate were it not for two processes which release it from the sea water. First, in certain conditions of temperature, salinity, acidity, and pressure calcium carbonate can be precipitated directly onto the sea bottom, mostly in tiny particles. Second, far more significant, almost all marine organisms extract dissolved calcium carbonate from its bicarbonate solution for use in building shells, other protective coverings, or skeletons.

At this point secondary cycles tie into the principal one. Locked in an oyster shell or a coral-head the calcium carbonate may be liberated by another animal which devours

> "Rain and wind patiently gnaw away at the deposits of calcium carbonate on the land, washing them into the sea in solution at the rate of 500 million metric tons each year."

or bores into it. Or the shell may sink into colder, high-pressure areas of the oceans that are calcium-poor, there to redissolve.

(Practically no calcium carbonate is to be found deeper than 16,000 feet.)

The calcium carbonate story possesses ramifications reaching beyond the creation of coral reefs. Free bicarbonate ions comprise one of the ocean's strongest buffering agents, reacting with other minerals on the sea-bed for the maintenance of relatively stable ratios and acidities—during such "acid" events, for example, as the eruption of an undersea volcano. Indeed, it was carbonates and other carbon-containing substances in the sea which combined 3 or 4 billion years ago to form those precursors of life—the first organic molecules—that in time developed the capability of ultilizing sunlight for the synthesis of their own food, in more time evolved senses and other specialized organs, in still more time evolved us! Here yet another carbon cycle became established.

Man plays two roles in the carbon cycles. Like any other animal, he consumes organic (carbon-containing) substances and exhales CO_2. But in recent years he has supplemented this "natural" contribution by large-scale conversions of pure carbon into CO_2 through industrial operations that burn the fossil fuels, coal and oil. In the past 50 years the percentage of CO_2 in the atmosphere has risen measurably. Probably the ocean will save man from his peril—again! The amount of calcium carbonate precipitated to the ocean floor is a function of the amount of CO_2 in the air. And so, if all else fails, the oceans can probably ensure an acceptable range.

Coral reef. Hard and soft coral, various types of fishes and worms, crustaceans of all varieties, and sponges abound in this submarine scene.

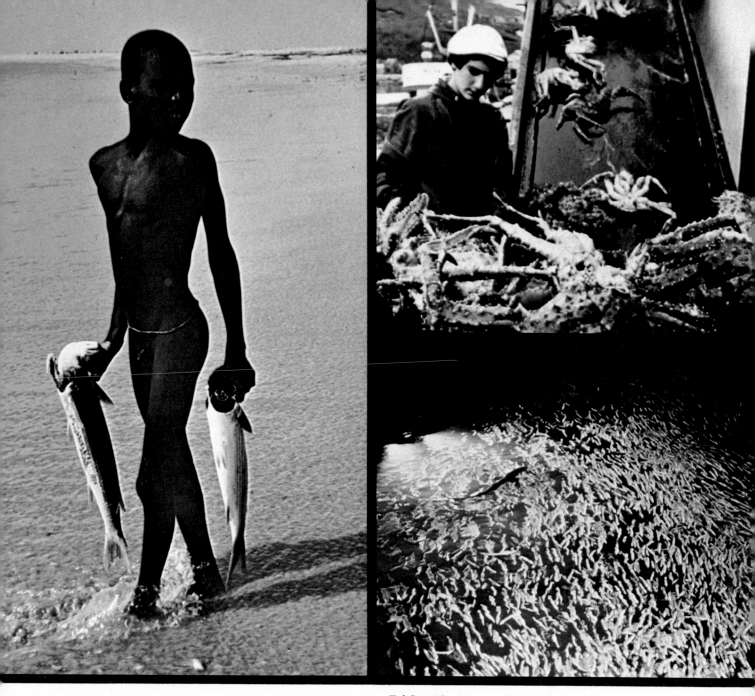

Fishing the Seas

A / Sport fisherman. *Strictly for recreation—although the fisherman can make a meal of his catch. Still, sportsmen must not deceive themselves into believing that the sea is inexhaustible.*

B / Spear fisherman. *Also a sportsman, armed to kill, but giving his prey more of a "chance."*

C / Native fisherman. *Dependent on the sea for sustenance, he takes only what he needs.*

D / Commercial fisherman. *Unfortunately, often by sheer negligence, these fishermen will endanger whole species.*

E / Squids attracted to light. *Commercial fishermen often use this method to attract fish.*

Chapter IX. Seasons and Provinces Of the Ocean

There are seasons in the sea, provinces as on land. And as on land, the sun's energy available at a given latitude together with the depth (or altitude) combine to produce the characteristic climates of the provinces. And in the oceans these are three-dimensional provinces, some of which interact with the land or the seabed, some of which are purely water provinces.

Today we barely have enough sounding profiles to outline a comprehensive geography of the sea. The continents are fringed

> "There are long, narrow abyssal trenches into which Mount Everest could be set and still have 6000 feet of water above its summit."

by littoral provinces, which extend underwater as the *continental shelf,* a vast province totalling 8 per cent of the surface of the oceans, ranging to the "drop-off-line" about 600 feet deep. A steep slope leads to the *continental margin,* then to the immense *abyssal plain,* 13,000 to 20,000-feet deep, dotted with volcanic islands and seamounts, streaked by mid-ocean ridges and rifts where the tremendous forces that push the continents apart are generated. Deeper are carved long, narrow abyssal trenches into which Mount Everest could be set and still have 6000 feet of water above its summit.

This general picture is made alive by rivers flowing from land, cutting canyons, pouring nutrients, pollutants, and sediments which trigger slumps and turbidity currents; by oceanic currents resulting from the density of the water and the rotation of the earth; by "vertical rivers" of cold water upwelling toward the surface; by constant intake of gases from the atmosphere and salts from the continents and the bottom; and by living creatures themselves, mainly plankton. Under the sea, the seasons can be as spectacular as on land.

The whole food chain responds to three dimensions as well as to a very high number of physical and chemical influences. Large commercial fishes like tuna, entire populations of sea birds, and marine mammals migrate in spectacular permanent exodus. For centuries these fantastic concentrations of sea life, achieved after millions of constructive years of unimpeded evolution, have enriched fishermen and gave birth to the myth of the inexhaustible mother sea. In fact, the average weight of organic matter to be found in a cubic mile of ocean water is to be counted in mere pounds, not tons.

A / *Sunset. The serene sea at nightfall.*

B / *Sea smoke. Warm air over cold water.*

C / *Fresh and salt water. A river flowing to the sea.*

D / *Spring. When springtime comes to cold Arctic waters, marine life blooms.*

E / *Bahamian trench. Islands in the tropics standing besides a deep abyssal trench (dark blue) as seen from a satellite photograph.*

Ocean Geography

The Atlantic. Like a spine up the middle of the Atlantic Ocean, the Mid-Atlantic Ridge snakes its way from the southern pole of the world to its northern. Probably it represents, geologists think, the crunch-line of gigantic crustal plates underlying the ocean floor that collide with each other as volcanic forces from earth's mantle set them in slow motion. Along the edges of the continents appear the continental shelves. Between Ridge and shelves are the ocean's abyssal plains.

The Pacific. The gigantic floor of the Pacific Ocean is laced and interlaced with ridges and fracture lines running at 90 degrees to the ridges. Scattered over the ocean plain are mountainous islands, sea mounts (mountains beneath the surface), atolls (islands sinking at the same rate that coral reefs are growing on them, forming the typical circular lagoons), and guyots (flattened sea mounts, also beneath the surface)—all souvenirs of the tremendous volcanic activity in this region.

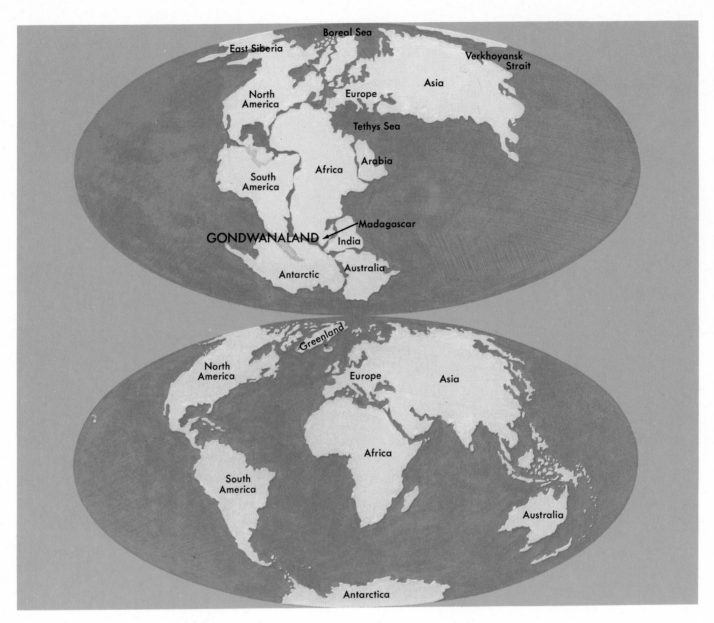

Continental Drift

In 1920 a German scientist, Alfred Wegener, published a theory he'd proposed for a number of years—that the land areas of earth had originally been one mass and that during the last 200 million years this supercontinent had broken apart and drifted to the degrees of longitude and latitude its parts now occupy. Today the theory of continental drift has gained acceptance. How do continents drift? To imagine such a thing, one must consider the solid land we inhabit as a light froth riding on earth's crust like bubbles riding in a pot of boiling liquid. In this pot the hot liquid moves to the surface and pushes the bubbles to the outer rim where cooler liquid descends. Essentially the same thing is happening to the earth. Heat from the core escapes along the central rift of the ocean and the upthrust material pushes the lighter continent apart.

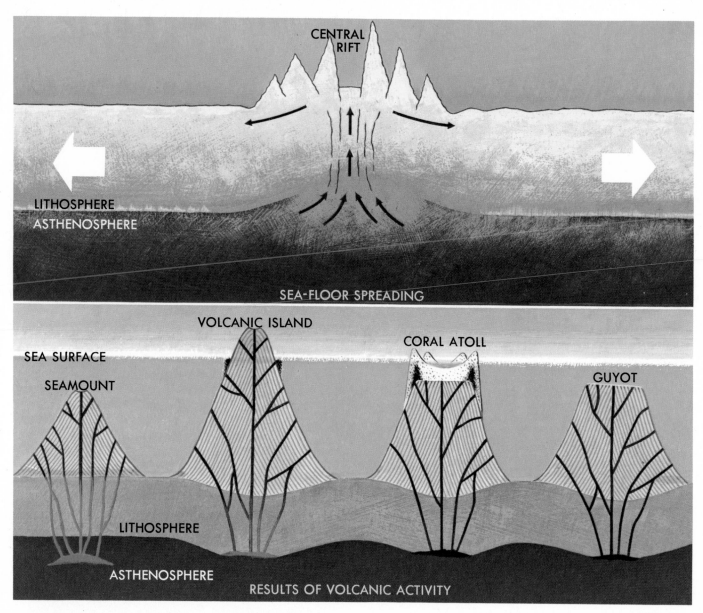

CENTRAL RIFT

LITHOSPHERE
ASTHENOSPHERE

SEA-FLOOR SPREADING

VOLCANIC ISLAND

CORAL ATOLL

SEA SURFACE

SEAMOUNT

GUYOT

LITHOSPHERE

ASTHENOSPHERE

RESULTS OF VOLCANIC ACTIVITY

Forces from Earth's Mantle

Islands of volcanic origin are found the world over. Among the most commonly referred to are the islands of the Pacific, especially the Hawaiian Island chain which extends for several thousands of miles. Some of these islands still readily display to the surface observer their volcanic origin, but others are more difficult to identify: for example, the coral atolls of the Western Pacific. But what is their origin? A popular theory today describes the earth's surface as made up of six large blocks, or crustal plates, similar to inverted saucers. Between these blocks we may find fracture zones which act as vents, permitting molten rock (basalt) to escape from the earth below the brittle crust. As this escape continues the submarine volcano grows upward, perhaps to a height sufficient to break through the surface, thus becoming an island.

Ocean Currents

The ocean's waters are constantly moving in a complex pattern of currents controlled by three forces: the wind, the rotation of the earth, and the differing densities of warm and cold water.

Winds blowing over the great expanse of open sea move the surface water in ripples, then in waves, and finally in swells. Currents move in the direction of the prevailing wind until the force of earth's rotation causes them to spiral off, to the right in the northern hemisphere, left in the southern.

Currents have an important role in regulating the world's climate by moving large masses of warm water to cold latitudes and cold water to warm ones. The climate in western Europe is far milder than that in similar latitudes because the Gulf Stream moves sun-warmed tropical water past its shores. Fog banks form off the coast of Southern California when warm surface waters are blown to sea.

Cold Arctic water sinks beneath the oncoming Gulf Stream and travels south in a countercurrent under it. This churning of the ocean's waters mixes them so effectively that the proportion to each other of dissolved salts is constant in all seas.

Speed of the currents varies, but it is estimated that the Gulf Stream moves at a maximum of six miles an hour as it sweeps past Florida and flows north. The cold countercurrent from the Arctic travels slowly along the bottom and may take several years to reach the Equator.

The Mediterranean Sea is almost landlocked, but it is still a part of the complex current system of the Atlantic. At the Straits of Gibraltar the Atlantic water flows east at a rate of about three miles an hour. In this arid region, sun and wind evaporate

WARM CURRENTS

great quantities of surface water, leaving dense salty water that sinks to the bottom where it is pushed along to the west in a countercurrent.

At the center of the whirling water of the North Atlantic lies a quiet, calm sea called the Sargasso Sea. Floating on this sea is a large meadow of seaweed, the *Sargassum.*

GREENLAND

ASIA

EUROPE

14

15

16

6

5

4

AFRICA

3

18

11

10

AUSTRALIA

19

9

20

21

1

� COOL CURRENTS

Global circulation: *Warm currents circulate clockwise north of the equator and counterclockwise south of the equator. Cold ocean currents do the opposite. The results include a marked effect on the climates of land areas adjacent to oceans. For example, the Gulf Stream tempers the climate of the American East Coast. And the California Current keeps places like northern California, Oregon, and Washington not too hot in summer and not too cold in winter. The differing water temperatures also act as barriers to some fish.*

1 ANTARCTIC WEST WIND DRIFT
2. PERU CURRENT (HUMBOLDT)
3. SOUTH EQUATORIAL CURRENT
4. EQUATORIAL COUNTER CURRENT
5. NORTH EQUATORIAL CURRENT
6. KUROSHIO
7. CALIFORNIA CURRENT
8. BRAZIL CURRENT
9. BENGUELA CURRENT
10. SOUTH EQUATORIAL CURRENT
11. GUINEA CURRENT

12. NORTH EQUATORIAL CURRENT
13. GULF STREAM
14. NORWEGIAN CURRENT
15. NORTH ATLANTIC CURRENT
16. CANARIES CURRENT
17. SARGASSO SEA
18. MONSOON DRIFT
    (SUMMER EAST, WINTER WEST)
19. MOZAMBIQUE CURRENT
20. WEST AUSTRALIAN CURRENT
21. EAST AUSTRALIAN CURRENT

Vertical Realms Of the Sea

Man swims through an ocean of his own —an ocean of air. He is as much a slave to it as any fish is to the ocean. I suppose most human beings are physically happiest around sea level. A healthy individual without cardiac difficulties can exist comfortably enough at the 8000-10,000 foot elevations of the Colorado mining camps. But at 17,000 feet, the altitude of some of the sky-high villages of the Andes, modifications have been made to the blood, the lung-capacities, and the hormonal system of the Indian inhabitants. The story is roughly the same in the ocean—with the reminder that water is 800 times denser than air and less transparent.

We look at the ocean depths in two ways: the "geography" of the earth underlying them and the "brightness" of the water spaces within them. First comes the tidal realm. Then come the continental shelf provinces — perhaps the most important area of the ocean, extending out to the drop of the continental slope where the true ocean begins. The shelf accounts for about 8 per cent of the earth's oceanic surface and goes out as far, approximately, as the depth of 600 feet. It is virtually another continent the size of Asia. At the margin of the con-

> "In the forlorn abyssal pit— 35,800 feet down—the crew of the *Trieste* found only a few odd animals an inch or two long. But they were still life!"

tinental shelf the continental slope steeply falls to the abyssal plains: immense undersea mountain-ranges and rifts, volcanoes sitting both above the surface (volcanic islands) or below it (seamounts), colossal trenches like those of the "ring of fire" around the Pacific Ocean many of which could easily swallow up Mount Everest and the Grand Canyon.

So far as light is concerned we divide the ocean into three zones: the sunlit zone, inhabited by plants and animals, which extends to about 600 feet or roughly the maximum depth of the continental shelf; the twilight zone, populated only by animals, from 600 feet to 4000 feet; and the black abyssal regions with their few and highly specialized animals. The bulk of life in the sea is concentrated in those waters that have some access to the sun: the sunlit surface waters inhabited by the countless constituents of the plankton, both plant and animal; the shallow seabed near the shores where stationary or crawling forms abound; the open sea just beneath the surface waters which is the domain of the oceanic fishes and mammals. And within the life of the ocean itself we also find the plankton layers and those mysterious deeper layers of small animals, rising and falling daily.

Perhaps the most interesting creatures of the sea are those which have at least partially liberated themselves from their "zone": the seal which can dive as far as 1000 feet; the dophin which can dive as deep and stay as long; the mighty sperm whale which can dive for one hour and reach the depth of nearly a mile. Beyond this depth life in the ocean thins out, and all is dark and silent. Almost all, I should say. When in 1960 Jacques Piccard and Don Walsh in the bathyscaph *Trieste* reached the bottom of the Mariana Trench, 35,800 feet deep, they found life even in that forlorn abyssal pit. Only a few odd animals, an inch or two long, subsisting on rare scraps of refuse which had drifted seven miles down—but still life!

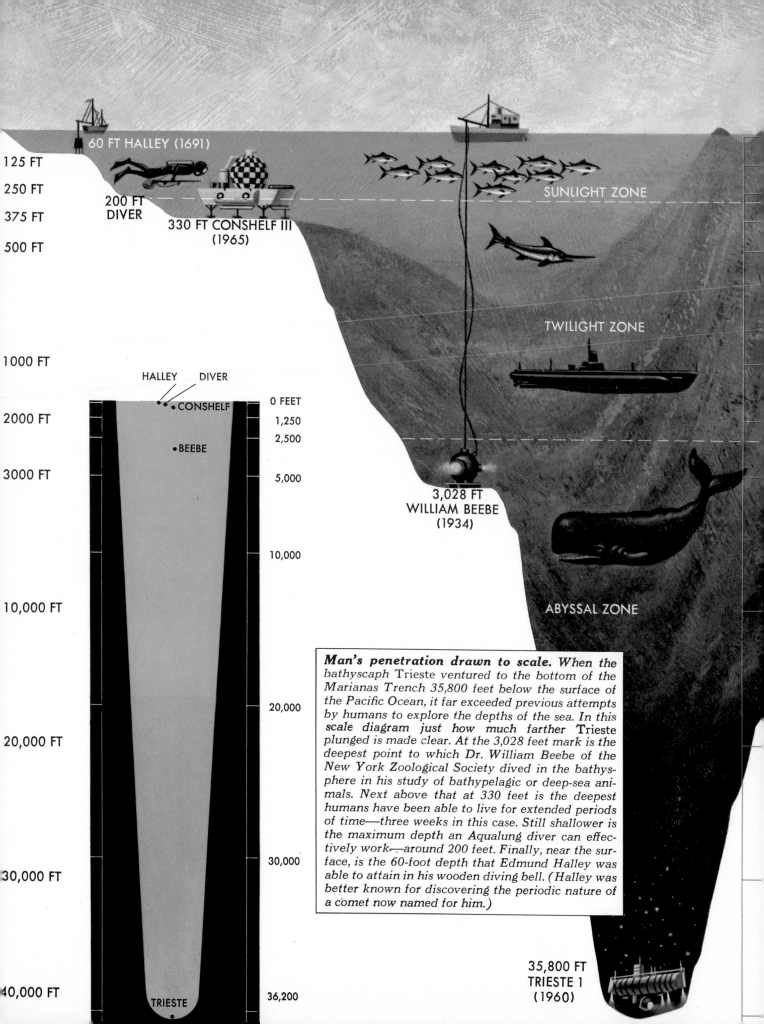

60 FT HALLEY (1691)

125 FT

250 FT

SUNLIGHT ZONE

200 FT
DIVER

375 FT

330 FT CONSHELF III
(1965)

500 FT

TWILIGHT ZONE

1000 FT

HALLEY   DIVER

2000 FT

• CONSHELF

0 FEET

1,250

2,500

• BEEBE

3000 FT

5,000

10,000 FT

3,028 FT
WILLIAM BEEBE
(1934)

10,000

ABYSSAL ZONE

20,000 FT

20,000

Man's penetration drawn to scale. *When the
bathyscaph Trieste ventured to the bottom of the
Marianas Trench 35,800 feet below the surface of
the Pacific Ocean, it far exceeded previous attempts
by humans to explore the depths of the sea. In this
scale diagram just how much farther Trieste
plunged is made clear. At the 3,028 feet mark is the
deepest point to which Dr. William Beebe of the
New York Zoological Society dived in the bathys-
phere in his study of bathypelagic or deep-sea animals. Next above that at 330 feet is the deepest
humans have been able to live for extended periods
of time—three weeks in this case. Still shallower is
the maximum depth an Aqualung diver can effectively work—around 200 feet. Finally, near the surface, is the 60-foot depth that Edmund Halley was
able to attain in his wooden diving bell. (Halley was
better known for discovering the periodic nature of
a comet now named for him.)*

30,000

30,000 FT

35,800 FT
TRIESTE 1
(1960)

40,000 FT

36,200

TRIESTE

Life In the Tidepools

Along the edges of the sea, where rocky outcrops form ridges and hollows, life abounds in tidepools that form when tides rise and fall, when surf ebbs and flows. These pools harbor algae and animals, the former providing food for the latter. And the latter feeding on each other sometimes. Mostly the animals are those that can evacuate the pools when they become dry on the outgoing tide—molluscs, fish, crustaceans.

A / Typical tidepool. *After the water has run out, algae, seaweed, and molluscs are revealed.*

B / File shell. *These slow-moving scallops have tentacles which extend from their shells and protect their eyes. The tentacles cannot be withdrawn completely.*

C / Rotifer. *Fast moving, microscopic, transparent creatures which occur in a variety of remarkable shapes. The largest of these glass-like creatures is about 1/50 of an inch long.*

D / Sea-urchins. *Brittle tentacled animals which move slowly. Some of the tropical species give off a*

100

poison. They range from one-and-one-half to ten inches in diameter.

E / Starfish and mussels. *Common starfish have five to ten arms or more. Most of them have spines with tiny pincers. The mouth is underneath the disc part of the body. Starfish feed off shellfish and destroy tons of them yearly. Mussels are prized as food and are sometimes as large as six inches long. They can, however, be poisonous.*

F / Grunts. *These noise-making fish average about a foot in length. Some can change color for camouflage.*

G / The return. *Water returning to the tidepool.*

H / Nudibranch. *Meaning "exposed gills," these shell-less molluscs are predatory animals, feeding occasionally on sponges.*

Life In the Coral Reefs

The reef community, in general, constitutes one of the oldest, most complex, and most tightly integrated of all ecosystems on this planet. Essentially, the reef is an offshore rock mass with a crest never much deeper than about 60 feet. Several types of reefs are found in the world's oceans: the fringing reef (in shallow water along coastlines), the barrier reef (running parallel to the shoreline), and the atoll (a circular island enclosing or partially enclosing a central lagoon).

A / Big-eye in coral overhang. Seaward of the back reef we come to the shallow reef flat where corals grow to a height equal to the level of the sea at low tide. The flat—with holes, shallow depressions, and small tunnels—provides an uncountable number of hiding places for small fish and crabs, sea urchins, sponges, sea fans, and fragile corals.

B / Shoal of herring. A reef is constituted of two or more zones, adding up to an environment which can sustain many different types of marine life.

C / Anemone. The back reef of the fringing reef system is a shallow lagoon-like area carpeted with sediments from broken skeletal systems of long-dead reef occupants. But we find occasional outcropping of patch reefs formed by isolated colonies of coral which break up the otherwise smooth reef.

D | Sea star. *The famous "crown of thorns," which not only feeds directly on the coral animals but is increasing in such numbers as to constitute a major ecological disturbance throughout the Pacific Ocean.*

E | Angelfish and clown anemones. *As we swim seaward over a reef we come to a buttress zone: the outer limit of the reef system. Here the ocean floor generally slants upward at an angle of 45 degrees or greater to form a wall against the mighty sea.*

F | Grunts. *Many fishes live some of their lives in the open ocean but more in the warm shelter of a great reef.*

G | Anemone. *These hungry, but delicate, animals can survive only in the protection of the reef wall.*

H | Yellowtail jacks on reef. *Because of the force unleashed by oceanic waves the buttress zone is made up of the more durable corals, including tree-like structures whose branches form in line with the waves.*

103

Life On the Continental Shelf

Along the edge of the world's continents is a shelf of land that extends into the sea to a depth of some 600 feet. On these continental shelves is concentrated a large portion of the plant and animal life of the sea. Here sun streams down and nourishes that which itself provides nourishment. And as algae flourish in these waters over the shelf areas, so does the animal life.

A / School of young barracuda. These large (five or six feet long) fish have been known to attack man. They feed by ripping with their teeth through a school of smaller fish and then returning to eat the pieces.

B / Sand dab. Both of this bottom fish's eyes are on its left side. It is not particularly large, rarely growing to longer than 15 inches, and weighing a maximum of two pounds. It has a large mouth which might indicate that it feeds on live prey which it surprises.

C / Birdfish. This agile, elusive fish lives in rocky reefs in the Indian and Pacific Oceans.

D / Reef shark. This killer, sometime weighing in at 250 pounds, feeds at night. It is believed to slaughter netted fish, and to lay waste to whole schools of bluefish.

E / School of mature fish. Like Bermuda chubs, these fish school near reefs on the continental shelf.

F / Sting ray. This is a flat fish with a long tail housing a sharp, poison-filled spine which it can swing in any direction if it feels threatened.

G / Diver in coral reef. Here a diver descends to the serenity of a coral reef and disturbs the local inhabitants.

H / Investigating. Divers are here looking over the wreckage of a World War II Japanese plane, on which marine growth has begun to take root.

Life In the Open Sea

Open sea—land of contrasts, contrasts of abundance and scarcity. The clear blue transparency of open ocean waters is the color of liquid desert just as sable gold is the color of the Sahara Desert. As in the Sahara, the desert of the open sea is patched by oases of life—huge schools of fish, sprats, sardines, flying fish. These fishes gather in the widely separated fertile areas of the open sea that abound with phytoplankton, the drifting microscopic plants on which they or those they prey on feed.

A | Sun on the open sea. Beneath the ocean's placid surface teem countless millions of sea creatures—some of which are described here.

B | Swordfish. These flat-billed fish swallow their victims whole. They will attack anything, often with no apparent purpose. Broken-off swords have even been found in the "skins" of whales and submarines!

C | Diatom. Here in the open sea, trillions and trillions of tons of plankton (both vegetable and animal) are produced and dispersed.

D / Shark. *Cruising the open sea at an average of 20 miles per hour, this predator seeks prey.*

E / School of fur seals. *Migrating from all points in the North Pacific to the Pribilof Islands off Alaska for breeding, these mammals feed on fish and squid.*

F / Whale. *A barnacle-encrusted whale surfaces to exhale and catch a fresh breath of air. These huge mammals often travel in "gams."*

G / Barracuda and jacks. *The barracuda is perfectly streamlined for cruising open water in mostly leisurely fashion. The jacks are built for sustained high-speed swimming.*

H / Bottle-nosed dolphins. *These intelligent mammals often travel in small family groups, sometimes in "pods" of hundreds.*

107

Life In the Deep Sea

In the deepest part of the ocean, known as the abyss, the water is ice-cold and the water pressure is greater than 1000 times atmospheric pressure. There is no light from the sun, but almost every animal there is bioluminescent, so that the abyss is alive with luminescent dots. It is not unlike the sky on a clear, dark night. The contours of the abyss are as varied as those of the continents, with steep mountains and mountain chains, valleys, and vast flat plains. Onto this primeval landscape of the ocean floor has fallen a steady rain of sediment. A study of these sediments reveals layer upon layer of different materials that have drifted downward onto the sea floor over hundreds of millions of years. Geologists often can translate this information into a history of our earth.

A / The "ring of fire." Here, on the ocean's bottom, volcanic lava helps to form the seascape.

B / The abyss. *Resembling the lunar landscape, this region goes as far down as seven miles! It differs from the moon's surface in one very important way, however—life exists here.*

C / Brittle-star and sea-urchin. *Even in the deepest parts of the ocean, these relatives of sea-stars carry on daily life in a light-less environment.*

D / Larval deep-sea perch. *Its enormous teeth, even at this early stage in its development, mark this fish as a predator to be reckoned with.*

E / Arrow crab. *This crab delicately picks minute animals off the deep-sea floor as it stalks about on eight spidery legs.*

F / Viperfish. *The hinged jaw of the viperfish allows it to engulf large fish whole as they float lifeless to the bottom. In addition, the roof of its mouth contains 350 photophores which create a beacon of light with which to attract shrimp and little fish.*

Birds Of the Sea

Within almost every group of animals are species which have returned to the sea. Birds are no exception. The sea is often a vast source of food. Various birds have adapted in a host of dimensions to life at sea. Webbed feet, elongated bills for burrowing in mud, and strong eyes are but a few ways birds have attuned themselves to their environment.

A / Adelie penguin. *One of only two species of penguin found regularly in Antarctica. They make long migrations through frigid waters for feeding and breeding.*

B / Horned puffin. *This uncommon bird has a remarkable triangular bill and is stubby and short-necked. It usually breeds from Maine to southern Greenland.*

C / Brandt's cormorant. *These large birds (body length 30-36 inches) sit upright and swim with their bills pointed upward. The Japanese keep cormorants to fish with!*

D / Flock of shore-birds. *Some birds at the edge*

of the sea find their food in the shallows and on the tidal flats.

E | Humboldt penguins. Living along the rocky cliffs of South America, these birds fish in a cold-water current that courses up the coast carrying with it many nutrients and fishes.

F | Guano birds. The source of a large Peruvian fertilizer industry, these birds feed on a fish called anchovetta which also provides Peruvian fishermen with their greatest resource.

G | Blue-faced booby. Found in the Caribbean, this two-foot-long bird makes spectacular plunges into the sea for its meals.

H | Double-crested cormorants. With its black body and orange throat, this is the only cormorant found on lakes and inland rivers.

111

Mammals Of the Sea

The sea is a difficult place to live for any warmblooded, air-breathing animal. Legs are useless, water quickly absorbs body heat, air can only be obtained at the surface, and the dense liquid medium is extremely difficult to travel through. In spite of this the development of an aquatic life style occurred a number of times in the geologic past and common features appear in differing groups.

A / Seals. *These animals are as happy on ground as in the water. Their short, thick fur has all-too-often been used to clothe people. But as we can see from these photographs it looks good on them too, so really should be left where it belongs.*

B / At home. *The seals are seen here on Seal Island in South Africa.*

C / From the Calypso. *After we had the seals on board with us for a few weeks, we could easily go diving and swimming with them.*

D / Elephant seal. *This baby seal is not waiting to feed. It is emitting a loud scream, which it does for*

112

attention and when angry. It weighs around 75 pounds at birth.

E / The battle. These two adult male elephant seals are fighting and screaming furiously in a battle over mating or feeding—probably the former. They have been known to weigh as much as 7000 pounds!

F / Manatee mother and baby. These vegetarians are docile, plain, and stupid. Their ancestors millions of years ago were elephants!

G / Bottle-nosed dolphin. Here a mother dolphin nuzzles her baby. These mammals can stay submerged for about 15 minutes.

H / Manatee and meal. In Florida manatees are protected and encouraged to clear the water-hyacinth-choked channels. It is one of the manatee's favorite meals as well as its calling.

Chapter X. The Pharaohs Of the Sea

The Great Pyramid at Giza, built by the Egyptian Pharaoh Cheops between 2613 and 2494 B.C., one of the largest structures on Earth, stands over 400 feet high and contains about 2,300,000 blocks of stone, each weighing over two and a half tons.

It is a pygmy structure compared to those erected by the "pharaohs of the sea," builders of reefs and atolls, principal among which in our day is the coral polyp. A coral reef is made of the limestone skeletons secreted by innumerable coral polyps and left behind after they die, one on top of another. In 1837 Charles Darwin visualized a coral reef as developing around the top of a volcanic mountain that was subsiding into the sea at just the right rate to keep the coral polyps housebuilding at their favorite level —the top 150 feet of the tropical sea. After World War II, engineers from the United States Navy tested the theory by driving a shaft down through layer after layer of ever-more-ancient coral at Eniwetok Atoll. And at 4222 feet the drill hit into volcanic rock (proving the theory).

Corals are the chief animal reef-builders today, but many other creatures are associ-

> "Polyps of colonial corals are all interconnected by lateral attachments—so the living coral colony lies entirely above its skeleton, completely covering it."

ated in reef-communities. In the past, other species, now extinct, played the lead role; they are responsible for the older sections of reefs. The reef association of plants and animals in the tropical waters of the world is the most complex of all ocean ecosystems. It is also the oldest ecosystem in the world, more than 2 billion years old.

Nine tenths of a typical reef consists of fine sandy detritus, cemented by animals and plants such as the enveloping plates of calcareous algae. Physical and biological processes convert this stabilized detritus into limestone. Remains of dead reef organisms make a contribution to the detritus. This major component of the reef has a fabric different from the upward growing lattice of stony algal deposits and intertwined coral skeletons that forms the reef core.

At the top of the modern "reef community" pyramid, the stony coral produces a calcium skeleton. Large colonies of polyps feed at night—waving minute food into the pharynx. The skeleton is composed of calcium carbonate crystals—secretions producing a skeletal cup called a theca within which the polyp is immovably fixed. Polyps of colonial corals are all interconnected by lateral attachments. So the living coral colony lies entirely above the skeleton, completely covering it. Polyps live in symbiosis with microscopic one-celled plants (zooxanthellae) embedded in the animals' tissue, where they are nourished by nitrogenous animal wastes and where they add oxygen to the surrounding water. These promote the calcium metabolism of corals.

The complex reef community consists of other members—limestone-secreting algae, limestone-secreting families of sponges, monocellular foraminifera, microscopic colonial animals, spiny sea urchins, echinoderms, bivalves, clams and oysters, all of whose accumulated skeletons and shells contribute to the reef limestones.

Coral atolls. As seen from this satellite photograph, these coral rings are surrounding volcanoes which once were above the surface of the water but have since collapsed and been submerged.

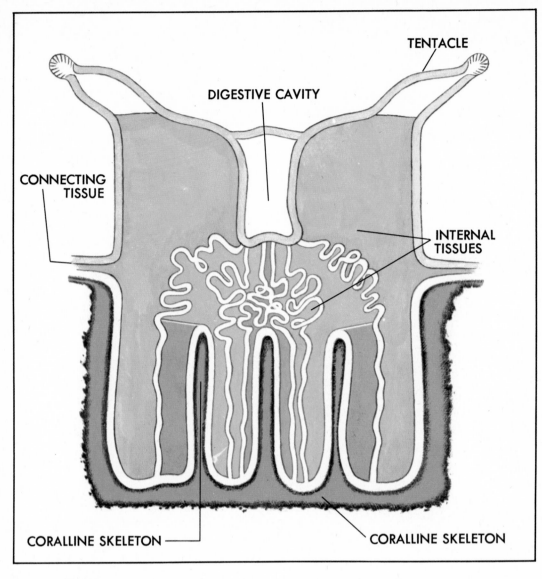

TENTACLE

DIGESTIVE CAVITY

CONNECTING TISSUE

INTERNAL TISSUES

CORALLINE SKELETON

CORALLINE SKELETON

The Coral Polyp

Most individual coral polyps are extremely small, but they are a colonial species and their colonies can attain considerable size. The animal can be thought of as a kind of self-service stomach. With its pharynx open to the sea, the coral waits for currents to wash something in the way of an edible object toward it. This the coral can receive passively or, using its tentacles, actively. In either case the food-particle is passed down into the central cavity of the animal, where it is ingested. What makes this humble little animal a "pharaoh" is that the lower part of the coral polyp's outer layer secretes a calcium carbonate skeleton, a cup called a theca within which the animal is immovably fixed. The polyps of the corals in a colony are all laterally interconnected—so that in a sense the colony itself, lying entirely above the skeleton and completely covering it, can be considered a single creature. The colony grows by means of the budding of new polyps. Countless numbers of these living, growing colonies have created the mountainous, infinitely variegated reefs we find everywhere in tropical seas.

A Stupendous Disaster?

Certainly two and maybe four times in the past 500-600 million years the reef community has "collapsed," with dozens of species of animals becoming extinct. After each of these collapses the reef community around the world has managed to reestablish itself with new balances, new species of animals in the typical "reef" association. Perhaps the most spectacular of these collapses was the fourth and, so far as we know, the last. This occurred about 65 million years ago at the close of the Cretaceous period. This was a period of great extinctions. Nearly a third of the families of all animals known in late Cretaceous times were no longer alive at the beginning of the next era, the Cenozoic. The reef community was not exempt. Two-thirds of the known genera of corals died out at this time, in addition to many other major groups. This late Cretaceous disaster also affected the land—this was the time of the end of the dinosaurs. Of the 115-odd genera of dinosaurs found in late Cretaceous fossil deposits none survived the end of the period.

To understand the reasons for such a stupendous disaster we must realize that the Earth has much changed its shape in these millions of years. Today's land area occupies about 30 per cent of Earth's surface. In the late Cretaceous almost two-thirds of today's land area was submerged under shallow oceans. The climate was warmer everywhere; there were no polar ice-caps; the weather of the whole world was of a piece. Then, at the end of the Cretaceous period, as a result of movements of Earth's crust, deep sea-basins were created which drained off the shallow oceans from much of the land. When the Antarctic ice-cap was formed some 20 million years ago, tremendous changes occurred in the climate, the weather, the currents of the oceans. What members of the reef community, including corals, were able to survive the long environmental change retreated to their present range: the tropical seas.

It is perfectly possible that there could be a fifth collapse of the reef communities. It would be accompanied by drastic changes in shorelines and weather. Man, with his technological ingenuity, could undoubtedly survive any geological vicissitude as a species. The only threats that could seriously endanger him are those that he would create himself. What happens today to the coral reefs all around the world is difficult to understand and it is too early to name it a new "collapse." But it is a very serious regression. The exuberant varieties of coral are disappearing from the northern and southern part of the Red Sea; from the Seychelles Islands to the Mozambique straits the fringing reefs are choked with sand and die. For several years most of the reefs in the Pacific Ocean have been badly damaged by a large coral-eating starfish, the "crown of thorns."

One thing is certain. If the coral reefs should disappear, the world would lose in variety and stability, and our grandchildren would

> "Certainly two and maybe four times in the past 500-600 million years the reef community has collapsed—with dozens of species of animals becoming extinct."

be deprived of the most beautiful sceneries the planet has ever produced. Who does not wish to swim or dive in the lagoons of the Caribbean or the Pacific—with gigantic brain corals standing in their gardens of sand, perhaps festooned with the mustard-hued "fire coral" on which every novice burns himself at least once.

Tales From the Maldives

It takes centuries for a coral reef to grow. The work is occasionally undone within a fraction of that time by man or nature. An example is found in the geographic history of the Maldive Islands, a group of atolls in the Indian Ocean west of the tip of the Indian subcontinent. Those islands, surrounded by coral reefs, were formerly a

> "Within a 24-hour period one starfish can graze over a reef area twice the size of its six- to 12-inch central disc."

bustling stop on the great spice routes of Europe's merchant princes. Today they are almost forgotten. Traditionally the islanders built their homes of coconut thatch. As the modern world began to catch up with the islands, large buildings and roads had to be built, and the only construction material available was that from the coral reef. Now the Maldivians harvest coral the year round, using large chunks of it as building blocks or breaking it up to make caulk. By destroying the fringing coral reefs, the Maldive people are unknowingly dooming their islands, for without a protective ring to shield the atolls from the pounding of trade-wind waves, the islands must themselves eventually disappear.

Thousands of miles to the east in the Pacific Ocean, Guam and other islands are similarly threatened. But this time man—the apprentice of the sorcerer—had to give an explanation. There had to be a scapegoat, and the animal indicated was the prickly starfish known as the crown-of-thorns. Even Australia's 1,250-mile-long Great Barrier Reef, the largest coral reef in the world, had been attacked by these creatures. Though it always was a member of the coral reef community, this predator of polyps was going through a mysterious population explosion. Today it has invaded the reefs of the Pacific, and one can easily spot its path, patches of bleached coral skeletons stripped of their tiny polyps. The dead coral is rapidly covered with a thick fuzz of algae that prevents new coral growth. Bereft of their living polyps, the reefs can easily be destroyed by the erosive action of the pounding sea.

But this is an oversimplified picture: within a 24-hour period one starfish can graze over a reef area twice the size of its 6-to-12 inch central disk. Of course, the solution is not for man to add one destruction on top of another and to attempt to eradicate the starfish population! Incidentally, if speared or cut in half, these crowns-of-thorns are capable of surviving and growing new arms. Islanders naively try to poison them individually or to collect them and bury them on land! Some scientists recommend importing their main natural enemy—a giant sea snail. One theory is that the population explosion is simply a natural phenomenon. In Australia it is believed that excessive collecting of the triton sea snail has permitted the starfish to reproduce abnormally. In Micronesia it is suggested that in the process of dredging channels, or dynamiting for fish, man has killed the coral or harmed it enough to alter the balance of the environment. Under normal conditions, most of the millions of eggs spawned by the crown-of-thorns are eaten by the live, healthy polyps.

Loss of the Maldive Islands. The coral reef protects the islands from strong winds and waves. In these photographs we see men unwittingly destroying their homes. They are taking huge pieces of coral from the water for use as building blocks.

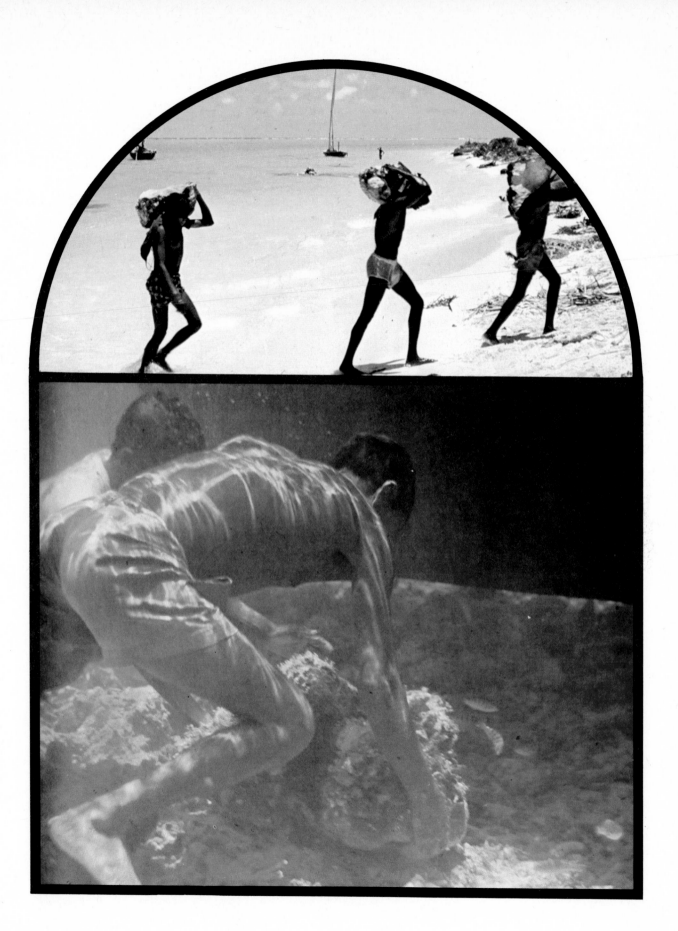

Chapter XI. Riches From the Sea

Energy is obviously earth's crucial resource. All the energy on earth springs from a single mother-lode, the sun: a gigantic thermo-nuclear furnace in which the fusion of lighter atoms into heavier ones releases energy which radiates across a gulf of 96 million miles, bathing the earth and all that live on it in light and heat.

The sun has been shining on the earth since the planet was born—some 5 or so billions of years ago. And for perhaps half that period the sun's rays have fed into "energy accumulators" on earth's surface: life-forms, plant and animal, which store up solar energies in their leaves and tissues. So for 3 or 4 billion years the sun has presided over countless generations of living things whose dead re-

> **"Sooner than we think we will have to reach out directly for the sun's energies, no doubt by the installation in space of orbiting stations to catch and concentrate its rays."**

mains have sunk into earth's crust—there, after many millions of years, to be converted into our "fossil fuels," coal and oil. From the giant rainforests of the Carboniferous Period, from the corpses of uncountable trillions of animals large and small, come all the coal and oil on earth, with much of the energy the sun poured into them still locked in, await-ing the key of fire. At last man entered, dig-ging holes in the ground, uncovering these caches of energy. He reared his new material-istic civilization on them. In 1900, 96 per cent of the world's industrial energy came from coal. Recently oil has challenged coal's dominance. But still today mankind's en-ergy needs are met mostly by the fossil-fuels.

In our time the fossil-fuel resources of the ocean have been called upon to supplement the fast-fading stocks on land—with large-scale drilling enterprises on the continental shelves and sea-beds. But like all bank-accounts the fossil-fuel energy bank-account has its limits. With his population booming from 1.5 billion in 1900 to 3.5 billion in 1970 to a probable 7 billion in the year 2000, a population which furthermore is increasing its per capita energy demands at an accelerating rate, man can already fore-see only too clearly the day when he has exhausted earth's fossil-fuel legacy terres-trial *and* oceanic. Fossil fuels, quite rapidly, must grow rarer and more costly to get at.

The sea-bottom also begins to be mined for various minerals—phosphates, heavy met-als, complex manganese nodules. Valuable metal-salts, used in the manufacture of drugs, are extracted from sea-water itself. And a wealth of new drugs, such as anti-biotics, can be isolated and developed in the ocean. Man must accomplish these extrac-tions of riches from the sea with more cau-tion and finesse than he is accustomed to. You cannot simply rape the ocean—at least, not indefinitely. Every mine, every factory, every processing-plant, every new construc-tion of any kind destroys to a degree the natural environment around it. The oceans' greatest food resource is of course fish, of which about 60 million tons are landed a year. There is no question but that the sea's yield of fish can be multiplied many times if blind fishing methods are progressively re-placed by scientific "farming."

The sea. "In our time the fossil-fuel resources of the ocean have been called upon to supplement fast-fading stocks on land."

Exploitation Of the Sea

A / Oil well. *In the near future almost 50 per cent of the world's oil will come from the continental shelf.*

B / Food fish. *Sophisticated fishing activities in Arctic waters will soon threaten a number of marine species.*

C / Manganese nodules. *This valuable mineral is about to be taken from the seabed for use in industry.*

D / Underwater oil well. *Oil companies presently employ more divers than any other marine industry.*

E / Kelp. *Kelp is necessary to a healthy marine environment but also useful to man in the manufacture of paint and other commodities. Kelp harvesters and ecologists are working together to reforest kelpbeds along the California coast.*

F / Sea-weed for food. *In Japan, particularly, seaweed is a popular addition to the everyday diet.*

G / Fish farming. Here concrete structures provide growing surfaces for algae and corals which attract fish populations for farming by men.

H / Scallop. Important as a food animal, the scallop is also valued for its attractive shells which are sold for use as plates or ash trays.

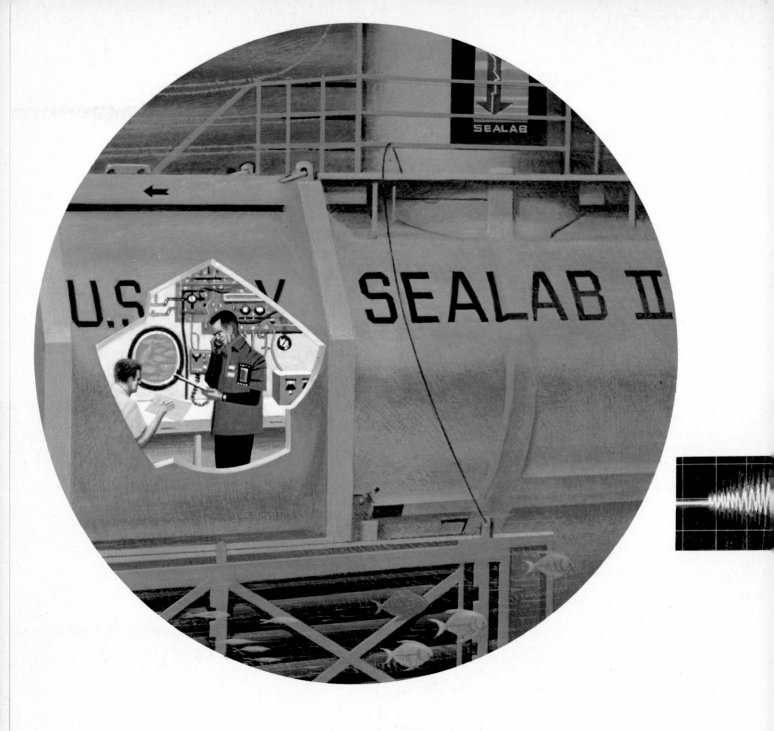

"Sea Lab to Conshelf: Are You There?"

In 1965 the first inter-undersea telephone conversation took place. Astronaut-Aquanaut Scott Carpenter from the United States Navy's Sealab II off La Jolla, California, spoke to French oceanaut Philippe Cousteau in Conshelf III off Cap Ferrat in the Mediterranean—6000 miles apart! The greetings exchanged between the two pioneering underwater stations were confused by static, language differences, and helium, the last of which causes the pitch of the voice to rise and become almost unintelligible. Now that man can live and work in the depths for weeks at a time, "saturation divers" have opened up millions of square miles of the continental shelf to man.

129

Endangered Species

One of the few animals who kills for reasons not involved in his own survival is man. We have carelessly and selfishly brought to the brink of extinction some of the most magnificent animals ever to inhabit the earth. We destroy animals for their pelts, or because they get in our way, or simply because they are there. Each animal species that disappears hastens the day when man himself will become an endangered species.

A / The polar bear. *One of the largest living carnivorous land mammals. Much larger than the females, males can weigh in considerable excess of 1000 pounds. Species exist only in and around the Arctic Ocean. There are probably 10,000 polar bears in existence, about half of which are in Canada.*

B / The Galapagos penguin. *This bird cannot fly, but is an expert diver and swimmer. Its population is numbered now at less than 2400.*

C / The dugong. *With manatee (illustrated here), this is the only surviving representative of a once much larger order. With whales and dolphins they have "returned to the sea."*

D / The blue whale. This is the largest animal ever to have existed. It can be up to 120 feet long and weigh over 135 tons. One marked blue whale is known to have travelled 1900 miles in 47 days. There are probably now fewer than 1000 of these animals left. (Pictured here is a humpbacked whale.)

E / The pelican. The insecticide DDT unintentionally affects this bird's ability to provide the calcium needed to give its shells rigidity. The real danger to this species is seen when we realize that not all pelicans mate every year, and those which do lay only two to four eggs.

F / The Juan Fernandez fur seal. A distinct species known to live only on the Juan Fernandez archipelago 500 miles west of Chile. Perhaps only a few hundred survive.

Chapter XIV. Oasis in Space

When Neil Armstrong's foot made contact with the surface of the Moon a new chapter opened in the story of the Animal World. For the first time an animal had consciously escaped from his own destiny. During all the millennia prior to Apollo 11 the propagation of species had been left to wind or water or to wings, fins, or legs. But in our own day here is man deciding on his own—and proving—that he can leave his ancestral environment—to the planet of his choice!

Our Solar System will prove to be no bargain so far as life is concerned. Scientists may have a field-day on Mars or one of Saturn's moons but none will feel homesick

> "Whenever man jumps across the interstellar void, and whatever our descendants discover among the billions of stars and systems that comprise the Universe, man will never find a planet that he loves as he does earth."

for them. And we cannot even guess how many generations or centuries will pass before man, armed with instruments of propulsion now only dreams, will make the next jump—across the interstellar void in search of new suns with habitable planets. But whenever this happens, and whatever our remote descendants discover among the billions of stars and systems that comprise the Universe, man will never find a planet he can love as he can earth. Nowhere else will the breezes of a mountain-meadow at dawn so perfectly suit his lungs. Nowhere else will gravity so perfectly suit his musculature. Nowhere else will a sun's rays so perfectly suit the pigments of his skin. And

nowhere else will an ocean so perfectly resonate with the salts of his own blood. Man shares this unique habitat with the whole community of the Animal World—all of whose members are our cousins in one degree or another. Scientists tell us it is a mistake to assume that animals think and feel like humans. But the dolphin's reactions are uncannily similar to man's. The dolphin is so fast and agile that he catches his food easily. He needs companionship and affection. Unable to live alone, he has a social life in the form of strictly organized herds. Delinquency is punished by ostracism. This is the most severe punishment, as isolation means demoralization and death. He falls in love like any human adolescent, but polygamy is the rule. He (or rather she) gives birth to live young after eleven months of gestation.

In our books to follow we will explore the grand themes of this ocean that man carries with him wherever he goes. We will try to show how the instincts and drives and senses and appetites of animals and plants link up and interweave with each other, creating the colorful, complex, constantly changing tapestry of life in the sea. We must realize that man can better understand himself by understanding the sea from which he sprang. We carry more along with us on our journey through time and space than earth's ocean. Our baggage includes our zoological heritage—a heritage which, as we twentieth-century specimens know only too well, can fall dangerously out-of-tune. If we are to survive as a species, we must reconcile with nature and the sea.

Man and dolphin. *Here one of* Calypso's *divers hitches a ride on the dorsal fin of his "cousin," the dolphin.*

Index

ILLUSTRATIONS AND CHARTS:

Sy and Dorothea Barlowe—15, 30, 31, 33, 68, 69, 81; Walter Hortens—17, 22, 59, 64, 95, 116; Howard Koslow—33, 54, 55, 63, 92, 93, 94, 96, 97, 99, 128, 129, 133, 136, 137

PHOTO CREDITS:

Chuck Allen—83; Black Star Publications: J. Launois—131, Flip Schulke—88, Pastner—125; John Boland—52, 70, 71, 72, 73; Fred Bruemmer—84, 85; Bruce C. Coleman, Inc.: R. Thompson—39, Jane Burton—77, Jen and Des Bartlett—73, R. Mariscal —28, 101, S. Gillsater—84, 138; David Doubilet—42, 46, 79; Kingsley C. Fairbridge—19, 91; C. P. Idyll—49, 124; G. L. Kooyman——, 35; L. A. Hyperion Sewerage Plant—28; Don Lusby, Jr.—47, 85; Marineland of Florida—13; Mobil Oil Corporation: M. Anguti—124; National Audubon Society: A. W. Ambler—42, Leonard Lee Rue, III—39, Karl Kenyon—40, 41, George Lower—77, John Gerrard—131, Molly Adams—88, N. Smythe—38, 39; NASA—11, 18, 91, 115; National Marine Fishing Service—29; NUC: Steve Leatherwood—66, 67; Naval Photographic Center—11, 19, 91; Osborn Laboratories of Marine Sciences: Walter Lerchenfeld—26, 52, 77, 100, 104; Harry Pederson—104; Photo Researchers: Russ Kinne—29, 77, 83, 101, 104, Peter David—108, 109, Fred Baldwin—106, Rudolph Freund—139, George Holton—138, Joe Munroe—49; Rapho—49; F. M. Roberts, © 1966—73, 83; Dr. Roman Vishniac—53, 70, 72, 100; WHOI—108, 124; © Douglas P. Wilson—29.